Erosion and Weathering

First English language edition published in 1998 by
New Holland (Publishers) Ltd
London - Cape Town - Sydney - Singapore

24 Nutford Place
London W1H 6DQ
United Kingdom

80 McKenzie Street
Cape Town 8001
South Africa

3/2 Aquatic Drive
Frenchs Forest, NSW 2086
Australia

First published in 1997 in The Netherlands as
Erosie en Verwering by Holding B. van Dobbenburgh bv,
Nieuwkoop, The Netherlands
Written by: A. Zevenhuizen
Translated from the Dutch by: K.M.M. Hudson-Brazenall

ISBN 1-85368-746-4

Editorial direction: D-Books International Publishing
Design: Meijster Design bv
Cover design: M.T. van Dobbenburgh

Reproduction by Unifoto International Pty, Ltd

Technical Production by D-Books International
Publishing/Agora United Graphic Services bv

Printed and bound in Spain by Egedsa, Sabadell

CONTENTS

Introduction 3

1 Erosion and Weathering 4
Energy 5
Rivers, rain and wind 7

2 Arches 17
Vertical faults 18
Resistance to erosion 19

3 Geological Cycle 28
Colours 28
Geological cycle 28
Catastrophism 33
4.5 billion years old 38

4 Weathering 45
Chemical weathering 46
Internal processes 54

5 Deserts 59
Subtropical deserts 61
Coastal deserts 64
Weathering and erosion in deserts 65
Sand deserts 68
Shield deserts 70

6 Caves 77

PHOTO CREDITS

Burrington, R. Photography, Denver, 23; Buys, A., 43; Corely, M., 11; D-Books International, 54[b]; Dobbenburgh, M.T. van, 5, 10[b], 15, 26, 27, 28[b], 29, 31, 33, 35[t], 37, 45, 46, 49; Dobbenburgh, B. van, 8, 28[t], 34, 36, 40, 44, 47, 55, 56, 57, 58; Eelman, E.J.J., 10[t], 53, 59; Fey, T./Foto Natura, 6; Foto Natura, 61; Harvey, M. Foto Natura, 75; Hopman, J./Foto Natura, 78; King, V., 7, 14, 22, 24, 30, 38[r], 39, 42[t]; Lemmens, F./Foto Natura, 13, 42[b], 60[b], 62[t], 63[t], 64, 65, 66, 67; Meinderts, W.A.M. /Foto Natura, 12, 50, 77; Meyvogel, J./Foto Natura, 77; National Park Service, Utah, 18; Pilon, J./Foto Natura, 63[b]; Sanders, A./Foto Natura, 32; Schwier, P.K, 48, 51, 52[t], 54[t], 68, 69, 70, 71, 72, 73, 74, 75[b]; Struik, S.A., 16, 20, 21, 79; Verwoerd, P./Foto Natura, 52[b]; Waard, A.J. de, 38[l];

r=right, l=left, t=top, b=bottom, c=centre

Introduction

Nowadays we all recognise the image of our planet taken from space, that beautiful, small, blue planet, with its oceans, continents and clouds. As far as we know, our planet is the only one with life, where water lies in great oceans, and with an oxygen-rich atmosphere as a protective layer around us. Clearly this planet is quite different to all the other, often drab planets in a vast lifeless space. The earth is a planet that is alive and constantly in motion with major cities, highways and aeroplanes. Not only are people, animals, clouds and rivers continually in motion, but also the planet itself, the mountains, the valleys and the plains are in perpetual motion. Major and minor changes are always taking place within and on the surface of our seemingly rigid and stable ball of stone. Whole continents are in motion, islands appear in the middle of oceans as a result of protracted volcanic eruptions and mountain chains such as the Alps are formed from what were previously ocean floor bedrocks.

Immense changes are continually taking place, but no-one has ever witnessed these happening, with the exception, now and again, of a volcanic eruption.

The reason why most of these changes make little impression upon us is due to the time scale involved. The time scale of days, months, years and even centuries that we use to order our lives and our history is rendered meaningless when attempting to describe the history of the earth. Physical changes often take hundreds of thousands or even millions of years to occur. The keyword to the earth is time. Of the many processes that have resulted in physical changes to the earth, the results of weathering and erosion described in this book are the most highly visible for us. Majestic and slow, like all processes of mother earth, they define the final shape of the earth on which we live.

1 Erosion and Weathering

The earth, in spite of its diameter of just over 13,000 kilometres, only consists of solid rock in its thin outermost layer, which is several tens of kilometres thick. Weathering and erosion are processes which result, respectively, in the breakdown and transportation of this solid rock. Although this sounds negative, weathering and erosion can be considered as the natural sculptors of the earth's surface. These processes create the fine detail in the crude forms of the earth's crust.

Weathering takes place wherever solid rock appears on the earth's surface and meets the elements, including air and water. Water penetrates into tiny cracks or spaces in a rock and from here, slowly but surely, starts breaking down the rock. Different forms of weathering may be active simultaneously. If the water freezes, it increases in volume and the fissures and hollows become larger. The water may also absorb certain elements from dead plants and animals, becoming slightly acidic and thus able to gradually dissolve a rock or parts of a rock. Sometimes the strength of plant roots is all that is required to widen these cracks. This process can be found at all levels from the microscopic to the macroscopic, and is responsible for loosening both tiny fragments and huge sections of rock.

Weathering happens all over the surface of the world, but whether it is a slow or fast process depends, in part, on the type of rock and the climate. One very important factor is whether or not any material that is loosened through weathering remains in the same position after being loosened. If weathered material remains *in situ* for too long, then no new material is exposed to the elements. If the weathered fragments of rock are transported away quickly and continuously, then weathering will be much more extensive and rapid. The removal of weathered material, or transport, is the work of erosion and occurs with the aid of the action of wind and water. The most important and most powerful manner of transporting weathered material is flowing water in a river. On a large scale the principle of erosion is simple: the higher sections of the earth's crust, above sea level, are broken down with the lower-lying sections, those below sea level, being filled. Sea level is the base level as far as erosion is concerned. Any water on the land above sea level does its utmost to get back to sea level. The higher it is above sea level, the faster it will flow downhill. In the same way as water powers a watermill, so it is able to transport weathered material, always downstream. Sometimes the flow of water is so powerful that it can transport huge boulders, but even if it is only carrying pebbles and gravel, through time, whole mountain ranges are flattened to sea level by the combined actions of weathering and erosion. The small-scale processes of weathering and erosion together have the power to make great changes in the earth's appearance, provided they act over a long enough period of time.

Fortunately there are other processes, the so-called internal processes that are at work in the innermost part of the earth, which produce new land and mountain chains, thrusting up these areas well above sea level. These processes produce coarse shapes in the earth's crust which are then broken down again by weathering and erosion. Thus there will always be high mountains and deep valleys and the earth will never become a smooth, uninteresting ball.

Energy

Just as we need energy to move and grow, the continuously changing earth also needs energy, amongst other things, for weathering and erosion. The earth has two huge sources of energy available to it, the sun's energy and the heat from within the earth.

The same source that supports most forms of life on earth, the sun, also supplies the processes of weathering and erosion with energy. The sun's warmth is responsible for the hydrological cycle. Water from the oceans evaporates and forms clouds that drift onshore. Above the land, these clouds lose some of their water load as rain and snow. Large quantities of water are released, especially above mountain ranges where the clouds have to rise. Once it reaches the ground, this water is responsible for weathering of solid rock through frost action and solution, and also for the transportation of any weathered material. The higher up in the mountains that water is released, the greater the transporting power of that water. The volume of water that is involved in the hydrological cycle is immense, and the influence of this cycle on weathering and erosion is enormous. Wind is another important means of transportation in the erosion process. Wind, too, is mainly caused by the sun's energy, which warms the different regions

Of the many processes that bring about physical changes on the earth, the results of weathering and erosion are the most visible. Majestic and slow, like all processes of Mother Earth, they determine the final shape of the earth's crust on which we live.

This valley is also the result of weathering and erosion, the processes which result in respectively, the break-up and transportation of rock. Wherever rock comes into contact with the elements, slowly but surely, it begins to break-up. Loose stones and gravel are then removed by wind and water.

of the earth to varying degrees, resulting in the creation of air currents.
An immense amount of energy is also available in the centre of the earth in the form of heat. The earth is, in fact, a hot ball that is only cooling down very slowly, because the majority of the heat lost is replaced by the decay of radioactive elements in the earth's core. Even after some 4.5 billion years of cooling, the temperature in the core is still estimated to be 5,000°C. This energy, in the form of heat, is called internal energy and is the driving force behind the many processes that are going on, unseen, in the earth's core. One result of these hidden processes is the elevation of large sections of the earth's crust. Areas of many thousand square kilometres may be elevated hundreds of metres or even several kilometres, whilst other regions are slowly but surely sinking. These processes of elevation and subsidence are very slow, at most a few millimetres per year, but if they continue for long enough they result in large scale changes.
The two sources of energy, the sun's energy and internal energy, collide, in a manner of speaking, on the earth's surface. Those regions of the earth elevated by internal energy are continually being broken down by weathering and erosion, the processes produced by external energy.

The Colorado Plateaus, USA, large plains raised several kilometres by internal processes in the earth, are gradually being worn down by the dual processes of weathering and erosion. The tremendous erosive force of rivers is plain to see whenever large or small rivers wind their way through solid rock. The rate of erosion caused by running water can be much faster than that of rain and wind.

The impressive and rugged landscape of the Fish River Canyon in Namibia, which, at 60 kilometres in length, is the second longest canyon in the world. The canyon is part of the extensive desert region of southern Africa where, due to scant water supply, little erosion is taking place at present.

Rivers, rain and wind

At any given time, large sections of the earth's crust will be in the process of elevation. Many areas that now lie high above sea level were, millions of years ago, much lower-lying. Prior to uplifting, these areas, quite often, lay below sea level, where they acquired thick layers of sediment, the product of erosion from areas which were at that time high above sea level. These sediments, like sand and clay, were laid down in horizontal layers and form rock formations which can often be hundreds of metres thick. The great weight of the overlying layers compressed these sediments into solid stone, like sandstone and shale.
In the course of time, a region like this also becomes elevated and a plateau is formed, a high plain above sea level. If this plateau forms the link between a mountain chain and the sea, then many streams will carry the run-off from the mountains across the plain and down to the sea. As the plateau is thrust continuously ever higher by geological forces, the uppermost layers of loose, unconsolidated sediments are easily washed away, exposing solid rock.

Vertical faults which divide the rock of Bryce Canyon (USA) into square blocks can weather easily. Weathering concentrates in the faults and finally transforms an entire region of sedimentary rocks into a landscape filled with columns and totem poles.

Changing circumstances are characteristic of the earth. The geographical position of continents, world temperature fluctuations and seasonal changes influence the processes of weathering and erosion. This steep-sided river valley in Oman is clearly the result of a period of time with conditions other than those that prevail at the time of the photo. It has been formed by running water in a past century or perhaps only in recent years.

In the same way that running water can drive a watermill, it is also able to transport weathered material. Sometimes the water is so powerful that it carries whole boulders down with it, but even if it only carries pebbles and gravel, in the end, whole mountains will actually be flattened, right down to sea level, by the combination of weathering and erosion. The small scale processes of weathering and erosion combined have the power to make large scale changes to the appearance of the earth's surface provided they take place over long enough time periods.

Freed of the loose sediment, the streams cut down more and more vigorously into this solid rock. The streams join up and form rivers, so that the higher the plateau is relative to sea level, the greater is the ability of the river to transport weathered material and to therefore cut even deeper. The impressive results of the erosive action of rivers can be seen, for example, in many canyons, where deep valleys have been carved out by meandering rivers, leaving the various layers of the sedimentary rocks exposed in the steep valley sides. Of course, water erosion by rivers is not the only way in which erosion takes place. The effect of river erosion is very great and on a global scale, rivers are responsible for most of the erosion, but erosion also takes place away from rivers, through the actions of rain, wind and other elements. In places where rivers are so powerful that they, more or less, take their own course across the plateaus, they only allow themselves to be diverted by large-scale structures in the layers of rock. In these situations,

A view of the Grand Canyon, USA. Differences in resistance to weathering and erosion of the many horizontal layers of rock are expressed in the steeper and more gently sloped parts of the canyon wall. The more resistant strata protrude from the wall and form steep rock faces or even terraces. The layers of rock on the horizon are certainly not the uppermost layer of the original plateau. Many hundreds of metres of rock used to overlie this layer, but they have been removed by weathering and erosion.

◁*Many regions that now lie high above sea level and which are being worn down by weathering and erosion, lay much lower, millions of years ago. Often they were below sea level where, before they were elevated, they were covered in thick layers of sediments, the weathered materials from regions that then lay high above sea level. These sediments, such as sand and clay are deposited in horizontal layers, both thick and thin, which form sedimentary strata hundreds of metres thick. The immense weight of the overlying layers causes the sediments to be compressed forming solid rocks like sandstone and shale.*

In the Sahara, the greatest desert in the world, very little water is available for weathering and erosion. Nevertheless, the disintegration and transport of rock still occurs and the structures that are left behind are no less impressive than those of other regions.

it is easier to observe the small-scale structures of the rock itself when erosion is caused by rain and wind, because weathering is then more important than erosion. Small faults and weak zones in the rock then determine the final shapes and beautiful formations are often the result. A fine example of this is Bryce Canyon in the Colorado Plateau in the United States, which is actually not a canyon. The thick layers of the rock formation are incised with vertical faults. These faults run in two sets, in two different directions and meet at an angle. They form a crosshatch pattern of faults which divide the rocks, and if viewed from above, are seen as large square blocks. This grid iron pattern of faults is

As a result of the growth of salt crystals in faults and cracks, larger and smaller fragments are broken off. Having been reduced to grains of sand, the material can then be transported by the wind over great distances.

repeatedly filled with water, which is slightly acidic and slowly dissolves small rock particles, freezing and expanding at night, causing the joints to deepen and widen. The weathered fragments and particles are carried away by wind and rain. This is how the pillars in the photograph on page eight are formed. These pillars, too, will eventually disappear as they simply tumble over or reduce in size. The weathering of the pillars will be slower than the weathering process in the joints, which retain water for longer periods.

△If the finest grains of sand are drawn high into the air by stormy winds, the desert rapidly changes into a shadowy, ghostly landscape. The temperature can fall by as much as 15 degrees within a few minutes because the particles in the air reflect the sunlight.

▷△▽In regions where the rock formation has been uplifted after deposition, without any other further deformation, then the layers of rock remain horizontal. The consequence of this is that the uppermost layers, which are the first to be affected by weathering and erosion, are also the youngest rocks.

The heat of the sun is responsible for the hydrological cycle. Water from the oceans evaporates and forms clouds that drift landwards. Looking out over the Valley of Desolation in South Africa it is easy to imagine that the ability of water to transport material increases higher up the mountain.

2 Arches

Arches, the geological structures that we will look at in this chapter, can occur in an area with the same geological origin as that of the canyons and pillars of Chapter One, namely a raised plateau.
Despite their obviously smaller dimensions, arches are certainly no less impressive than canyons and pillars. Arches are arcs of rock that arise as the result of weathering and erosion by rain and wind. The most important factor in the creation of arches is the presence of deep vertical faults in a rock formation, the faults all running in the same direction and therefore more or less parallel. These faults repeatedly fill up with water and the combination of various weathering processes ensures that they slowly but surely become ever deeper and wider. The action of frost pushes the faults and cracks slightly wider apart every night and highly diluted acid, absorbed by the water from the atmosphere, slowly dissolves the lime cementing between the grains of sand, loosening them until they are washed or blown away. In the cracks, the destructive force of the water is far greater than on the surface of the rock and these cracks can, in the course of time, grow to become deep, parallel gullies several metres apart. The continuous removal, by wind and water, of the weathered products that have gathered in the gullies is very important if these gullies are to continue to deepen. This allows the continual exposure of new rock to the rigours of air and water permitting optimal weathering at the foot of the gully. Between the gullies, walls of rock, or the so-called fins, remain standing. As the gullies widen, these fins get smaller.
A rock formation is not usually homogenous. It often consists of many layers, both thick and thin, which are different in composition and react differently to weathering since they are of different resistances.
If the gullies cut into a softer, less resistant layer of rock, then this will easily weather down quickly, not just vertically but also laterally. When the removal of material takes place on both sides of the fin, a hole or window that connects two gullies is created at the base of a fin. Once this hole has been created, water and wind will continue with their weathering action and enlarge the hole so much that finally an arch is formed. In regions where it is frequently windy, the sharp edges and points of the fins and the edges of the opening will quickly be rounded due to sand blasting. The erosive action of the wind is at its greatest a few centimetres above the land surface as this is the area in which the wind can carry the greatest quantity of particles. Arches are, therefore, mostly the result of rain and wind erosion rather than by river erosion. They are, in geological terms, rapidly changing phenomena. They are often formed and erode away in less than 200,000 years, despite being the result of relatively slow weathering and erosion processes. In geological terms this process is rapid and it is, of course, due to the small-scale nature of arches compared to other geological phenomena.

Vertical faults

The parallel, vertical faults, which are the most important prerequisite for the formation of fins and arches, can arise in various ways. In general, they are the result of elongated concave and convex structures in a rock formation, called folds. The folding of a rock formation can be imagined as the rippling-up of many layers of carpeting laid in a room that is long enough but not wide enough. Elongated structures arise in the length of the room (or the longitudinal direction of a rock formation), along which parallel, vertical faults can occur. Often the folds result from internal processes in the earth, which deform or compress a rock formation in a

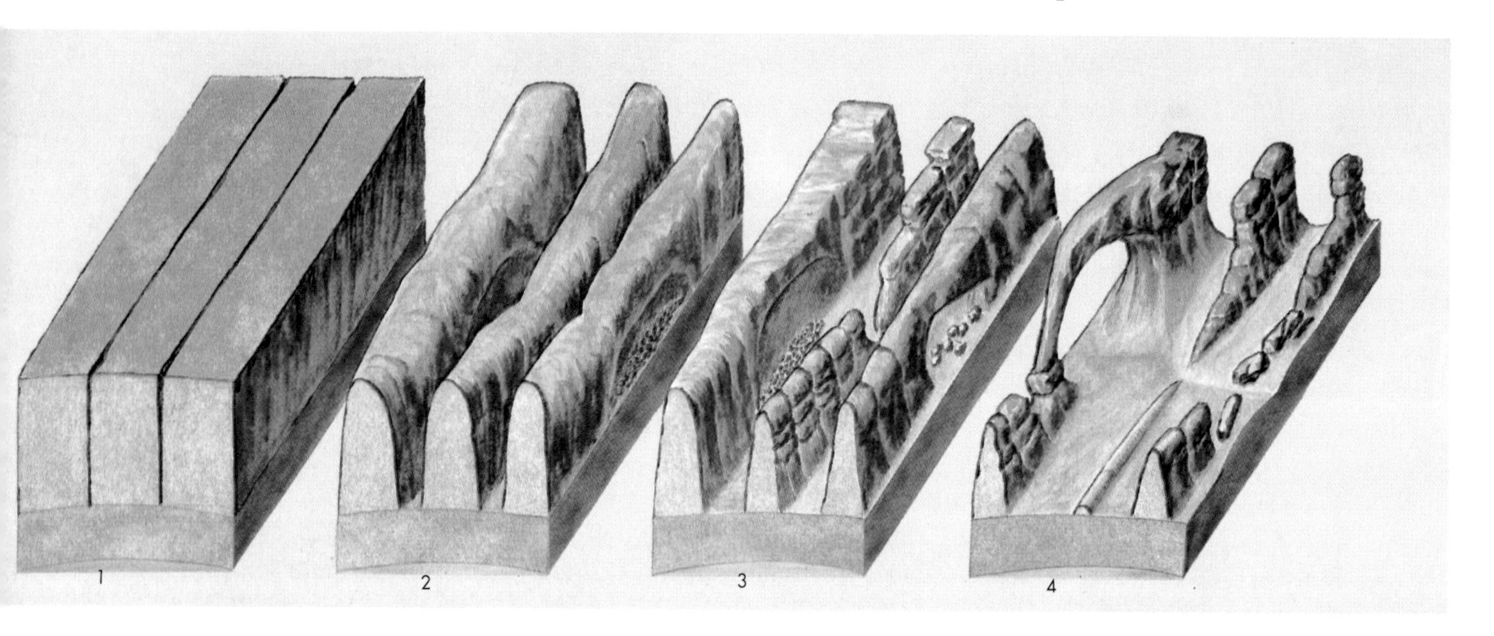

How are arches formed?
1. As the earth was thrust upwards, deep cracks penetrated to the buried sandstone layer.
2. Erosion wore away exposed rock layers and enlarged the surface cracks, isolating narrow sandstone walls or fins.
3. Alternating freezing and thawing caused crumbling and flaking of the porous sandstone and eventually cut through some of the fins.
4. The resulting holes were enlarged to arch proportions by rockfalls and weathering.

Arches eventually collapse, leaving only buttresses that in time will erode. Some natural bridges may look like arches, but they form in the path of streams that wear away and penetrate the rock. Pothole Arches are formed by chemical weathering as water collects in natural depressions and eventually cuts through to the layer below.

particular direction. Sometimes elongated, convex structures are the result of the unusual composition of the rock formation, without any involvement of internal processes, for instance, because a thick layer of salt underlies the rock formation.

Salt layers occur through the evaporation of sea water. This happens when a shallow sea, in which sediments have been laid down over a long period of time, is shut off from the ocean. If it is then only flooded with fresh oceanic water a few times every thousand years, the inland sea will dry up each time and the salt from the water will be deposited in thick layers. If this continues for long enough, then layers of salt will be deposited which are tens or even hundreds of metres thick.

If the circumstances then change, other sorts of sediments can be deposited on top of the layers of salt. An unusual characteristic of such layers of salt is that at a certain pressure they become unstable. The thicker the layers of sediment overlying the salt, the greater the internal pressure that will force the salt, at some point, to give way and flow to places where the pressure is lower. This usually means it squeezes upwards through faults and weak zones in the overlying rock. This creates spherical salt domes in the rock formation, where the salt has pushed its way through the rock formation at certain points. A similar process leads to the formation of the elongated convex structures that result in parallel, vertical faults.

Resistance to erosion

It is remarkable that all the structures we have examined display a stepped construction from the base of the structures upwards. In both large and small scale objects, steep sections such as canyon walls alternate with sloping sections. Sometimes terraces may even be formed. Arches are not always uniform in shape, but in places have narrower sections which appear to have been constricted. These narrower sections result from differences in the resistance to erosion of the different horizontal rock layers.
During the creation of rock formations, a process which takes millions of

years, the circumstances under which deposition took place may have altered many times. When large quantities of sediment from higher altitudes predominated, high levels of sand was deposited. Amongst these sand deposits are the skeletons of dead animals which lived in the shallow sea and sank to the bottom when they died. The calcium skeletons later formed the cement between grains of sand when the layer is compressed and compacted to form sandstone. If the rate of sedimentation is low, then the proportion of animal skeletons between the grains of sand increases and

Detail of an arch. The lower resistance of a layer of rock at the base of the fin means that the gullies on both sides are expanding rapidly. It is clear to see that the hole that has thus arisen is the result of weathering of both sides of the fin.

Vertical wall of sandstone, which includes the Wolfberg Arch, in the Cederberg Wilderness Area in South Africa. The wall (fin) has arisen in a horizontally stratified rock formation as a result of deep parallel vertical faults. The faults fill up time and again with water, and the combined action of weathering processes ensures that they slowly become deeper and wider.
The opening which later forms the arch, arises if the gullies at the base of the fin become much broader in a less resistant layer. The hole becomes larger and larger, whilst the gullies cut down further.

▷Detail of the Wolfberg Arch. The lowest layer of the hole, the five metre thick layer of sandstone which ends suddenly, is the non-resistant layer of rock in which the hole originally formed. The rock both above and below is clearly more resistant. The fact that the hole has expanded upwards, is as a consequence of the collapse of the overlying rocks. Definitely not a very safe place to stand!

Arches are often found grouped together in a region due to the unusual way in which arches occur, in which parallel vertical faults play an important role. Moreover, they often run parallel to each other.

▷Delicate Arch, in the Arches National Park, certainly lives up to its name. Once part of continuous layers in a rock formation, this giant is defying the forces of nature, but how long will this last ... What is noticeable is the asymetrical construction from bottom to top, with narrower sections in places. This construction is the result of differences in resistance to weathering of the mostly horizontal layers of rock.

the composition of the rock changes. If the sedimentation rate is very low, then layers of limestone are deposited. As we have already seen, in special circumstances even whole layers of salt may be formed. All of these sedimentary layers differ in composition. If many millions of years later, this rock formation is elevated and, for instance, is affected by the actions of frost and acid rain, then some layers will weather more easily than others. In the canyon walls and arches, the resistant layers therefore protrude more prominently in relation to the less resistant layers.

This enormous arch in the Wilderness Area of South Africa will probably collapse long before the slow processes of weathering and erosion have managed, grain by grain, to transport it away. Sometimes, geological changes take place at a faster pace, with large pieces of rock breaking loose. Arches are, geologically speaking, phenomena that change rapidly despite being the result of slow weathering and erosion processes by rain and wind. They are formed and broken down again within a period of less than 200,000 years.

The light coloured, gently rounded, resistant layer of rock on top protects large sections of these fins. Wherever the resistant layer of rock has disappeared, large sections of the fins have been removed. Differences in resistance of various rock layers is due to changes in the circumstances during the deposition of the rock formation. If many millions of years after the rocks were laid down, a formation like this is elevated, then one layer will weather faster than another. Arches National Park (USA).

Double Arch in Arches National Park (USA) actually consists of two arches one behind the other. The largest of the two is more than 50 metres wide and 30 metres high. The coloured stripes in the rock surface are not the rock layers, for here they lie horizontally. These stripes are the result of wind-borne fine clay particles, that contain iron and manganese oxide (so-called desert varnish).

Skyline Arch in Arches National Park (USA).
The opening in Skyline Arch suddenly doubled in size in November 1940 when a large block of rock fell out.

3 Geological Cycle

The pedestal or mushroom rocks are some of nature's most astounding, seemingly impossible structures. They are so fragile that even without the rigours of weathering and erosion they do not appear to be able to survive for long. However they are the result of erosive processes. Originally, mushroom rocks were a section of a continuous layer in the rock formation; these boulders continue to exist simply because they were more resistant, for whatever reason, than the rest of the layer. The less resistant layer of rock, just beneath the balanced stone is partly protected by the block of rock above it, whilst it is only the section of rock in direct contact with the stone which is completely protected. The stone and the rock layer beneath it are still continuously attacked by weathering processes, which explains the round shape of the stone. As the layer of rock under the stone weathers and erodes more quickly, so it decreases in size and strength and can no longer support the rock top, which eventually topples to the ground.

Colours

Most of the structures we have seen to date are remarkable for the magnificent colours they display; indeed, some of them appear to have been painted by the hand of a giant artist, so surrealistic are the colours.
Neat bands of colour can sometimes be followed across an entire landscape. Successive layers of different coloured rock are usually due to unusual minerals which are present in the layers of rock. Metals in rocks oxidise when they are exposed to contact with oxygen, for example, after weathering. Thus red and brown sandstone layers are often the result of iron oxidising to the mineral haematite. The iron is found between the sand grains in the lime cement. Yellow rocks can be due to iron oxidising to limonite and purple layers are often related to manganese oxide. In some instances, colours often appear later on rocks as if by an invisible brush, but in fact this is the work of the wind. Desert varnish, for instance, gives rock in dry regions a deep, dark, sun-bronzed colour. This varnish actually consists of very tiny clay particles containing iron and manganese oxides, which are wind-blown and coat the entire structure. This form of colouring has nothing to do with the layers of rocks and their internal composition, but can be found almost anywhere. Freshly weathered and eroded surfaces are not covered with desert varnish and tend to be lighter in colour than older surfaces.

Geological cycle

If weathering and erosion take place over long enough time periods, then little will remain of an elevated region. Whole chains of mountains will be flattened, and raised plateaus will eventually erode down to sea-level. The rate at which this happens slows down in the final stages. Weathering, frost action and the solution of rocks all continue, but river erosion reduces or ceases completely as the speed of the water flow decreases. Sometimes, if the land has been flattened to such an extent that the speed of the river is no

Balanced Rock in Arches National Park (USA), is more than forty metres high. The thickness of the non-resistant layer directly under the boulder is the result of this huge boulder rolling back and forth.

Just like Balanced Rock, the Queen Nefertiti Rock has been sculpted by nature in Jurassic sandstone. The red colour of this sandstone layer is due to oxidation of iron into the mineral haematite. Arches National Park (USA).

Cederberg, South Africa. Wherever weathered material is continuously removed, the surface of the solid rock is continuously exposed to the elements. Rainwater lies on the rock surface and continually refills the splits and cracks, allowing the various weathering processes to work optimally.

▷If weathering and erosion continue for long enough, then even thick rock formations can be eroded away. Whole mountains can be broken up this way and elevated plateaus flattened to sea level. Relicts of the unusual formations and structures remain visible for a long time. The most resistant pillars and fins, or at least parts of them, remain standing the longest. Whether this fin on the photos will remain standing for much longer is questionable. Arches National Park (USA).

longer fast enough for the transportation of sediment, then the sediment will be deposited by the river. The entire region will then only be subject to wind and rain erosion and will flatten out very slowly.

Relicts of the exceptional formations and structures that we have seen earlier remain in the landscape for a very long time. The most resistant fins and pillars, or parts of them, remain for the longest time period.

The processes and results of uplift, weathering, erosion (transport) and renewed uplift together form the so-called geological cycle. In the 17th and 18th centuries it was still thought that structures and landscapes, such as

Blue Mountains, New South Wales, Australia.
This photo demonstrates the two aspects of weathering and erosion. On the one hand, the small scale effect, where the rock is being broken up millimetre by millimetre, shows faults and cracks are clearly visible in the rock surface in which weathering begins. Both small and

mountains, valleys, canyons and deserts were the result of several major disasters in the 'short' history of the earth (so-called catastrophism). It was Hutton, a Scotsman, who at the end of the 18th century was one of the first to recognise, in rocks from his own local area, the processes of the geological cycle. Not only did Hutton observe that rocks are weathered, eroded and deposited again, but also that within rock that was in the process of being weathered and eroded, structures were visible that indicated that these rocks had been formed long ago in similar circumstances to the way rocks were being formed at the present time. Hutton knew nothing about the internal forces in the earth and could not

large fragments are removed at an agonizingly slow rate from the solid rock.
On the other hand, large scale weathering and erosion ultimately form the appearance of landscapes all over the world.

explain the uplift of regions. However, he produced an important theory which would confirm him as the father of geology, namely the theory of uniformity.

All geological processes that shape and change the appearance of the earth's crust have done so throughout the entire history of the world in the same way. For every old rock that we find, there is a geological environment in which, in our time, the same sort of stone is being formed. If we want to know how a certain rock is formed, we only have to go back in time with the present geological processes. In this way, the present is the key to the past.

Catastrophism

The theory of uniformity makes it clear that the present form of the earth's crust is almost entirely the result of known creative and destructive geological processes.

However, occasionally great catastrophes have also clearly left their mark on the history of the earth. Hurricanes and floods can have a major influence on the environment, destroying in just a few seconds what would normally take geological processes thousands of years to create. One of the largest disasters in the earth's history happened about 66 million years ago. A meteorite with a diameter of several kilometres hit a spot, as yet undiscovered, somewhere on earth. The resultant huge bang threw so much dust into the atmosphere that the earth was thrown into darkness for a considerable length of time, with far-reaching consequences for the world's climate. These changes took place so quickly that plants and animals had no opportunity to adapt, with the result that about 80% of all living animals at that time became extinct, including the dinosaurs. Only smaller animals were able to survive the difficult conditions.

Catastrophes of this size have happened throughout the long history of the earth and some were perhaps even worse than the one of 66 million years ago. Clearly, such catastrophes have had an effect on the earth's geology, because

If one considers the forces that were necessary to create such large, sinister structures, one can understand how many of these structures were, in former times, associated with evil forces. Arches National Park (USA).

Like so many unusual formations in the landscape, pedestal or mushroom rocks are the result of differences in resistance. What were once part of a continuous layer of rock in a rock formation have been left behind as the topmost blocks as a result of their high resistance. Arches National Park (USA).

◁Through wind and weather 'The Three Gossips' keep a watchful eye on everything and everyone in the Arches National Park (USA).

Soil formation is a self-sustaining process, as long as the soil is continually vegetated.

In the 17th and 18th centuries it was thought that all structures and landscapes, like mountains and valleys, canyons and deserts were the result of a few huge catastrophes in the 'short' history of the earth (catastrophism). Towards the end of the 18th century a Scotsman, called Hutton, was one of the first to recognise, from rocks in his own locality, several processes in the geological cycle. The geological cycle is the cycle of uplift, weathering, transport and renewed uplift.

▷*Straight, angular structures are often a sign of rapid weathering of a highly faulted rock formation. The faults constitute weak zones in the rock and weathering is much more rapid here than on the surface of the rock. Large fragments loosen and fall down. Surface processes are not given time to round off the angles thus created. The ultimate shape of the structures is, in fact, determined by the weak zones that were already present in the rock.* Arches National Park (USA).

Balancing rock in Yunnan Province, China. One of nature's fragile sculptures, which seems as if, even without the destructive forces of weathering and erosion, it may not survive for long. Yet this structure is also a product of these two processes.

▷△▷Hottentot people from the surrounding area were convinced of the supernatural powers of this balancing rock formation and called it 'The Finger of God'. It seems that the supernatural powers have abandoned it and left the fragile structure to earthly forces. These forces caused it to topple from its pedestal in 1993. Namibia.

climate changes influence the processes of weathering and erosion, but the effect of such catastrophes on life on earth was even greater. Imagine how the earth would be if 66 million years ago no meteorite had struck the earth. Would man rule the earth or would huge monsters be in charge...?
Disasters like that can, of course, happen any time; so there is still hope for the ants!

4.5 billion years old

An important consequence of Hutton's theory is the perception that the earth must be very, very old. Processes of weathering, erosion, deposition and elevation are so slow that, if the present complex formations and structures of the earth's crust are the results of these processes, then they must have been going on for a very long time.
The geological cycle does not allow us to determine the age of the earth exactly. Attempts have been made to calculate the earth's age from the total thickness of sediments and the present rate of sediment deposition. This method of calculating the age of the earth is not precise because large sections of rock formations have disappeared as a result of weathering and erosion. Furthermore, the rate of deposition of sediments has not been constant during the entire history of the earth and in some periods of the earth's history no sediments were deposited at all.
The discovery of radioactivity in 1896 means that we are now able to calculate the absolute age of the earth and the age of various rocks. This calculation is not easy, and many assumptions have to be made, but scientists nowadays agree that the earth is approximately 4.5 billion years old.

This balancing rock is also not immune to the geological cycle. It seems quite likely that the weathering and transport sections of the cycle will occur in the near future. Arches National Park (USA).

Angular forms of a rock have a relatively larger surface area than rounded forms, and in these instances, weathering concentrates on the corners. Highly rounded forms are the result of weathering of solid rock.

Wave Rock, Hyden, West Australia. This 15 metre high wave formation is caused by weathering over a long period of time.

The transition from one layer of rock to another is not always abrupt. Often the different layers of rock are the result of gradually changing sedimentation circumstances and the transition from one rock layer to another is thus also gradual. If, after a period of deposition, a period without deposition follows, then soil can form in the uppermost centimetres of the layer of rock. Soil consists of erosion materials, micro-organisms and plant remains, and gives the top layer of rock a dark colour. The longer this goes on, the deeper the soil that is formed, and the greater the depth of this dark colour. If, after a period of soil formation, the circumstances change and sediment is once again laid down, then this dark soil layer is preserved within the rock formation. The underside of the layer shows a gradual colour change, whereas the upper side changes colour abruptly.

4 Weathering

Weathering and erosion are responsible for the final configuration of the earth's surface and the landscapes as we know them.
The fact that there are so many different types of landscape on earth is, amongst other things, due to both the type and degree of weathering and erosion which take place. Weathering occurs throughout the world wherever rock comes into contact with the atmosphere and the hydrosphere.
Physical weathering, which includes frost action, usually begins with the formation of small faults and weak zones as a result of the reduction or release of pressure in the rock. Pressure release occurs when erosion removes thick layers of rock, so that the pressure on the underlying rock is greatly reduced. As a consequence these underlying layers expand and small cracks appear. The effects of frost action and the growth of plant roots gradually expand these small cracks.
Frost action is particularly common in regions where, at certain times of year, the temperature fluctuates around freezing point. When water freezes to form ice, it increases in volume by about 9%. For frost action to occur, water must be present as well as ice. This is because the ice attracts water

Fiery Furnace in Arches National Park (USA). Hundreds of flame-shaped columns and fins appear to 'burn' at the end of a warm day when the heat absorbed from the sun is radiated and a warm glow from the sun hangs low over the landscape.

and this causes the ice crystals to grow in size, which increases the pressure in the cracks and splits the rock. In regions with temperatures well below freezing, any water present only occurs as ice, and weathering through frost action is much slower than in regions where the temperature fluctuates around zero.

In addition to frost action, the growth of salt crystals can also expand fissures and splits in the rock. Salts that precipitate out of ground water, for instance, will grow in the same way as ice crystals and can exert great force on the rock, at times even cracking rocks open. Weathering through crystal growth occurs particularly in dry regions where many salts precipitate out due to high evaporation rates.

As rock formations react differently to pressure reduction, weak zones often arise between the layers. The layers or formations of layers become more or less detached along the layer boundary. New layers are repeatedly removed by weathering and erosion.

Not all stratification in rocks is the result of the original deposition in

sedimentary rocks. High pressure and temperatures (conditions that prevail deep inside the earth) can cause an originally homogenous rock formation to be metamorphosed into a layered rock, that when weathered, produces similar layered structures.

Chemical weathering

Chemical weathering is not simply the solution of limestone by acid water. In many rocks, other elements are present which are soluble in ordinary water, provided that they are immersed for a long enough time period. Often, certain rock minerals are chemically altered with the aid of infiltrating water to form other types of mineral, for instance, clay minerals. Generally, clay minerals are larger in volume than the minerals they replace. The result of this transformation to clay minerals is that the rock is peeled like an onion. Weathering occurs over the entire surface area and whole

▽▷Kings Canyon in the Watarrka National Park in Australia. An impressive river valley with a depth of up to 200 metres. The rock seems to drip down the steep sides. On the upper section of the canyon is 'The Lost City', a 'town' of cupolas, sculpted out of the horizontal layers of sandstone by weathering and erosion.

layers of rock peel off. This process is called exfoliation. Whether chunks, layers or grains come loose, the effect of all this weathering is to increase the surface area of the rock. As a result of weathering, the surface area of a rock that comes into contact with the elements is increased, which in turn increases the speed of weathering. Thus, weathering enhances its own actions.

However, during the course of time, if the weathered material is not removed by erosion, then the weathering will not go any deeper into the rock. Fresh rock is no longer exposed to the elements, whilst the weathered material itself is broken down into smaller and smaller fragments.

The particular form of weathering which prevails for any given rock type in any given region depends on various factors. One of the most important factors is climate. Climate is concerned with precipitation, humidity,

In regions with an extreme climate, weathering and erosion are often extreme, producing spetacular rock formations such as this one in Iceland.

temperature, wind and vegetation, and all of these components can all have an important influence on weathering and erosion. The actual combination of these components will determine which form of weathering or erosion is most favoured and will have the greatest influence on the final outcome. Chemical weathering for instance, can, in a warm climate with high precipitation and rich vegetation, quickly break down or alter a thick layer of limestone. Whilst limestone is almost always insoluble in ordinary water, water that contains a small quantity of carbonic acid can easily absorb calcium carbonate, the main constituent of limestone. Carbonic acid derives from the atmosphere and the vegetation. Rain water on its way through the atmosphere absorbs very small amounts of carbon dioxide and when infiltrating the soil also absorbs small amounts of carbon dioxide from plant remains. Carbon dioxide and water together form carbonic acid, highly diluted of course, but the higher the rainfall and the more luxuriant the vegetation, the stronger the acid becomes. Rain water infiltrates the soil and combines with the ground water that permeates through the limestone. The acid in rain water and in ground water dissolves large sections of the limestone underground, resulting in the creation of limestone caves and karst landscapes.

A karst landscape is one dominated by the products of chemical weathering. As a result of the solution of the rock by acid water, the faults that dissect a limestone layer are rapidly deepened and widened. Over time, towers of limestone are left behind between these gullies. If these gullies then fill up with sediments that are not liable to rapid solution, the landscape becomes more level, but dotted with the remnants of limestone towers. The material that fills the gullies between the limestone towers is not weathered material from the region itself, for this will all have disappeared in solution. Once again this process only happens in a warm, wet climate with lots of vegetation. This form of weathering can break down a limestone region much more quickly than any other forms of weathering.

In this rock formation on Tenerife, it is obvious that the thick layers, which clearly differ in composition, are also themselves composed of many layers of rock. These thin layers demonstrate slight changes in the conditions during deposition. The boundaries between the thin layers are gradual, but those between the thick layers are quite abrupt.

◁▽These ghostly limestone rocks in the province of Yunnan, China, are mainly the result of chemical weathering. Rain water on its way through the atmosphere absorbs minute amounts of carbon dioxide and when it then infiltrates into the ground it absorbs even more carbon dioxide from rotting vegetation. Together, carbon dioxide and water form highly diluted carbonic acid, but the greater the precipitation and the more luxuriant the vegetation, the stronger this acid solution will be.
In a warm, humid climate weakly acidic water can easily dissolve calcium carbonate, the main component of limestone. The faults that crisscross a limestone formation are rapidly deepened and widened by the acid solution. After a period of time only pinnacles of limestone remain.

Chemical weathering in a warm, humid climate with lots of vegetation can break up a thick limestone formation much quicker than other forms of weathering. If the faults in the limestone formation are deeply weathered then only small fragments of limestone remain between the gullies. Sometimes the lower-lying parts are filled with sediments that are not susceptible to rapid solution. This can create a relatively flat area with remnants of limestone pinnacles.

Apart from being responsible for extraordinary, yet often attractive, formations in the landscape, chemical weathering can also have an economic significance. Enrichment of the extractable mineral held in ore bodies can occur as a result of chemical weathering. Differences in solubility of minerals in a rock mean that the only minerals left, after long-term chemical weathering, will be those that are the most difficult to dissolve. Metal ores like iron and aluminium (bauxite) ores are very difficult to dissolve and are relatively enriched when other minerals have been leached out through being dissolved in water. Naturally this process

The results of the pressures and movements that occur in the earth's crust through plate movements are present in many places and at all sorts of scales, for example, as folds in a rock formation.

As a consequence of chemical weathering, certain elements and minerals, which are difficult to dissolve, become enriched in a layer of rock. Examples of such elements are metals like iron and manganese. The metals oxidise when they come into contact with oxygen from the air. This can produce some wonderful colours.

An unusual result of chemical reactions in rocks is the petrifaction of wood. If wood is rapidly covered with rock through which water is able to circulate, then rock elements which are in solution in the water are exchanged for elements in the wood. The wood is thus transformed entirely into stone, whilst all the original forms and structures in the wood, like year rings, are preserved. When the overlying rock is then removed by erosion, trees or stumps of stone are left over.
Above: petrified forest in yellow limestone. Cape Bridge Water, Victoria, Australia.
Below: petrified tree stump. Petrified Forest, Arizona, USA.

only works in regions where chemical weathering is more likely than physical weathering.
Another spectacular result of chemical reactions in rocks is the petrifaction of wood. If wood is rapidly covered by rock in which water circulates, then rock particles dissolved in the water can be exchanged for elements of the wood, a process called substitution. The wood is, in this way, entirely turned to stone whilst retaining its original structure and form, for example, its year rings.

Internal processes

As previously stated, only the outermost layer of the earth is actually solid rock. This crust is only a few tens of kilometres thick and is the result of a long period of cooling of the viscous ball that originally comprised the earth. The largest section of the earth's crust, roughly 90%, therefore consists of igneous rock, which is molten rock that has solidified close to the surface, for example, granite.
Igneous rocks are continuously being broken down by the processes of weathering and erosion, wherever these rocks come directly into contact

with air and water. After several billion years of weathering and erosion the surface of the earth (not the entire crust) now consists of approximately 75% sediments and sedimentary rocks.
The igneous rocks are really the source of most of the sediments. The composition of the sediments and sedimentary rocks is mostly dependent upon the composition of the igneous rocks from which they are derived. Igneous rocks can be very varied in their composition, depending on the combination of circumstances that have played a role in their formation. They are composed of a number of different minerals like quartz and feldspar.
Minerals, in igneous rocks, that are the least soluble form the largest proportion of the sediments on the surface. One of the minerals most resistant to chemical weathering is quartz, a mineral that is present in large quantities in continental igneous rocks. Clay minerals, that are formed through chemical alteration of other minerals found in igneous rocks, are also very resistant. The vast majority of sedimentary materials in the geological cycle thus consist of quartz particles (sand grains) and clay particles.

The dark layer is solidified lava flow, the result of a volcanic eruption, on which other sediments were later deposited. As warm, molten rock has a larger volume than cold, solid rock, a lava flow will shrink when it cools. The result of this is the occurrence of characteristic, vertical shrinkage, jointing. A homogenous rock layer is thus transformed into a colonnade, that from above, looks like a honeycomb.
The layer of volcanic rock is clearly more resistant than the other rocks and protrudes from the wall of rock.

Gigantic elephant's feet, formed by weathering and erosion. Weathering is concentrated in the deep gullies in the hill between the 'feet'. Water streams down the gullies off the hill and thus weathering is much more rapid there than on the projecting parts. These protruding parts will therefore be preserved for a long time. Arizona (USA).

The earth's crust, on which both the continents and the oceans lie, is not a huge single mass, but consists of various, large tectonic plates that move in relation to each other. For example, the African plate lies under the continent of Africa, the Eurasian plate lies beneath Europe and Asia and the Pacific plate is beneath the Pacific Ocean.
In the long history of the earth these tectonic plates have moved great distances. On at least one occasion in the earth's history, or possibly more often, all the continents have been joined together as one large super-continent. After this latest super-continent, called Pangaea, broke up about 200 million years ago, the plates carrying the continents started moving and finally located to their present positions. The plates are still moving today but so slowly that we do not notice it.
One of the results of the movements in the earth's crust is that pressures build up in the plates, for example where two plates push against each other or slide past each other. The forces that are involved in this process are so massive that whole pieces of the earth's crust are crumpled up or slide over each other. In this way, entire mountain ranges can appear at the edge of a plate, for example, the Italian Alps. Earthquakes are also a frequent consequence of such movements.
The results of the pressure and movements, which occur in the earth' s crust through plate movements, are present in many places and at many different scales, as, for example, folds in a rock formation. Such structures are, however, only visible once weathering and erosion have done their work and laid bare the structures.
At certain places, including plate boundaries, hot, molten rock comes to the surface in the form of volcanic eruptions. Lava flows spread out over the land. The much lower surface temperature, compared to the inner core, quickly cools the molten rock that solidifies. Sometimes the molten rock does not reach the surface, but remains trapped somewhere in the earth's crust. It may then force its way between two layers of existing rock, if this requires less force than going straight through the rock. This so-called intrusion forms a new layer of rock that did not exist at the time the rock formation was originally laid down. As molten, warm rock has a greater volume than cold, solid rock, a lava flow or an intrusive body will shrink when it cools. Characteristic, vertical shrinkage joints are the result. A homogenous layer of rock is thus transformed into a colonnade, which from above looks rather like a honeycomb.

5 Deserts

Extreme forms of both weathering and erosion occur within deserts because of the major influence of the climate. Although extreme, these forms are not uncommon. In fact, deserts cover approximately 25% of the land surface of the earth. Not all deserts are vast tracts of sand. The term desert applies to those areas where the annual potential evaporation is higher than the total annual precipitation, or where the total annual precipitation is less than 250 millimetres. A desert does not therefore have a defined appearance or a defined geological history. The climatic belts, which are mainly responsible for the location of deserts, occupy a more or less fixed position on the earth with respect to the North and South Poles. The continents that float on large plates on the earth's crust, move independently of the climatic belts and consequently anywhere on these continents could have a desert climate. Other factors influencing desert formation are latitude and the position of the land mass in relation to ocean currents.

Detail of a sand dune, one of the many beautiful results of weathering and erosion in extremely arid regions. About 25% of the surface area of the earth is covered in deserts.

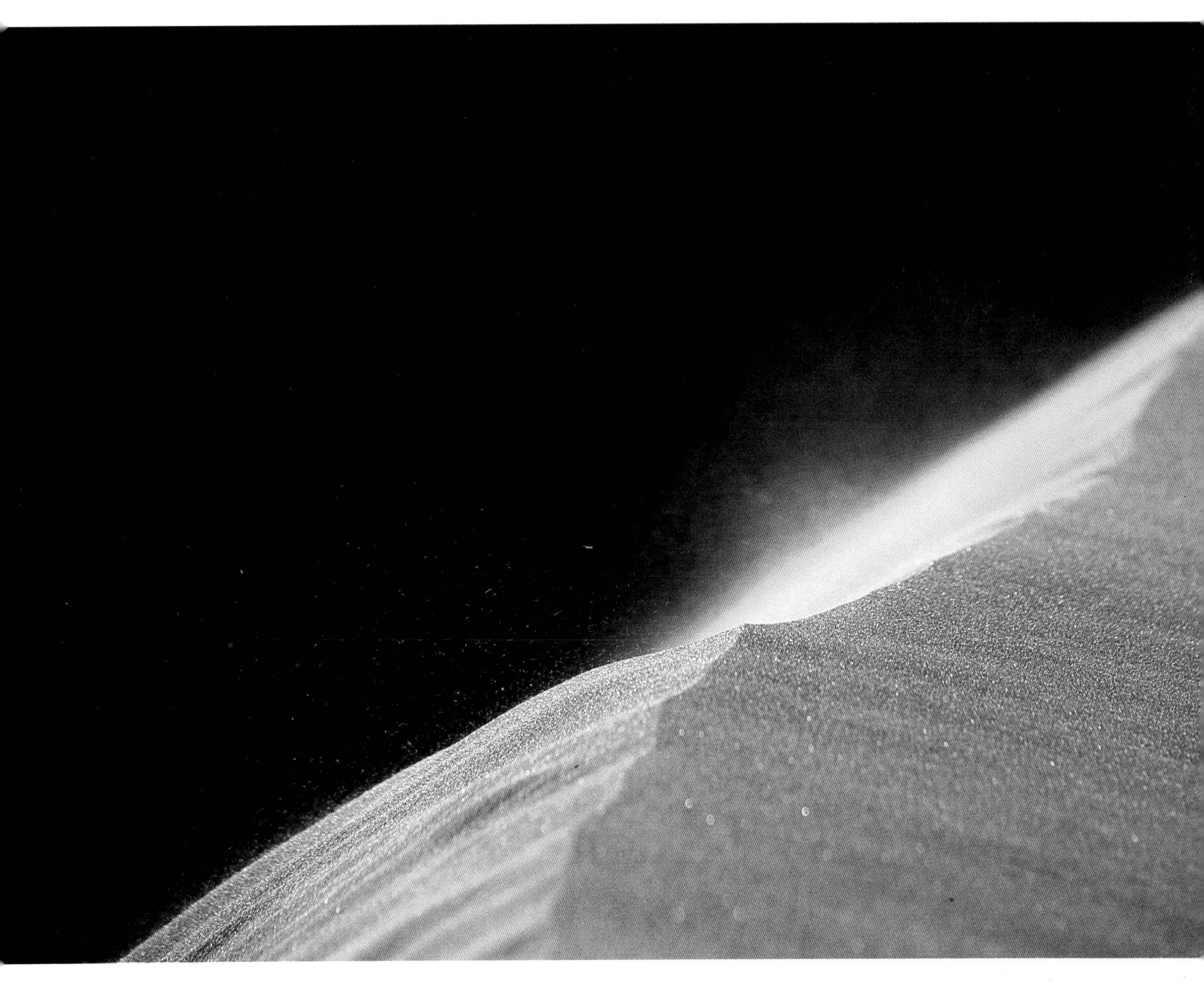

Sand deserts, those endless vast plains with nothing more than dunes and fata morganas only constitute a small proportion of all deserts on earth. Only about one third of the Arabian peninsula, the most sandy of the desert regions and one fifth of the Sahara, the largest desert in the world are actually sand deserts.

The commonest dunes are barchan dunes. They form diagonally to the wind with the rounded side to windward. As a result of the way in which dunes are formed, they migrate downwind at speeds of 10 to 30 metres per year.

Subtropical deserts

There are many different kinds of desert, but the image of a sea of sand that the word desert conjures up, naturally remains the most impressive. Most deserts belong to the so-called subtropical deserts, found in subtropical regions, where high temperatures prevail all year round. These are found mostly between 20 and 30 degrees north and south of the equator. These types of deserts are the result of a world-wide descending hot air current. Examples of these are the Sahara and the Kalahari Deserts in Africa and the Great Australian Desert. In addition to the extreme temperatures, which are capable of evaporating more than ten times the level of precipitation, subtropical deserts are characterised by low humidity and large differences in temperature between day and night. This latter characteristic is mainly due to the absence of clouds and a lack of vegetation. This means that about 90% of the sun's heat reaches the surface of the desert, that is thus highly heated during the day. By comparison, in temperate regions, about 60% of the sun's heat is reflected back into space by clouds and dust. At night, the desert surface cools rapidly because there are no clouds or vegetation to

Despite the fact that there are many different dune forms, the internal structure is, on the whole, the same. As can be seen on the photograph, a dune has a gentle slope and a steep slope. The gently sloping side lies to windward. The continuous deposition of sand grains on this side causes this side to grow both in height and downwind. The grains of sand that finish up lying near the top, make the top unstable and the top repeatedly collapses down the leeward side. The windward side grows, the leeward side collapses, thus the dune migrates in the direction of the wind.

Due to the extreme aridness, most of the forms of weathering and erosion that define the shape of the earth's crust in the rest of the world have no opportunity to be active in the desert.
Yet weathering and erosion does occur, even though the processes are much slower than in less arid regions.

retain the heat. The differences in temperature between day and night can, in subtropical deserts, be as much as 50°C. Precipitation may not fall for several years, whilst a single, rare cloudburst may produce several times the annual average rainfall at one time. Floods and mud flows are often a consequence of this, because the soil surface cannot cope with large quantities of water. In mountainous desert regions the water gathers in large basins, called playas, at the foot of the mountains, forming playa lakes. It may stand here for several weeks, but because of the high temperatures the basin will dry out through evaporation long before a new supply of water arrives. The salts that are dissolved in the water, are precipitated out on the floor of the playa, so building up thick layers of salts.

Sand dunes are formed when the wind blowing over the desert is disrupted by an obstacle, for example, a stone. Immediately behind the obstacle, the speed of the wind drops and the wind loses part of its load. Sand dunes can sometimes reach a height of 500 metres. Heights between 30 and 100 metres are quite normal.

Due to lack of vegetation in many deserts, there is a continuous stream of fine particles being transported by the wind. The finest particles remain in suspension a few centimetres above the ground, the somewhat larger particles jump a little further each time, landing on the surface and bumping into other particles, which in turn are bumped into the air. The wind's greatest erosive force is at a few centimetres above the land surface.

Coastal deserts

Another type of desert is the coastal desert. These are found in dry regions adjacent to the sea or the ocean, where shortage and excess of water form a stark contrast.
The cause of droughts in such a region is the combination of a subtropical high pressure area and a cold ocean current along the coast.

Ventifacts are characteristic of wind erosion in desert regions. Stones which project partly or wholly above the surface of the desert are sandblasted continuously. If the wind almost always blows from one prevailing direction, then a sloping side is the eventual result.

As a result of the rotation of the earth and the influence of the wind on surface currents, in some places on the earth, the oceanic currents flow away from the continents, allowing cold water to well up from the deep oceans to replace them. This cold water cools the air above the ocean, so that the temperature of the air drops and very little sea water evaporates. The result is dry air above both the ocean and the adjacent land.
Coastal deserts are noticeably cooler than subtropical deserts. Due to the proximity of the ocean, both the summers and the winters are temperate.

One characteristic of such desert regions is morning mist which occurs as a result of water vapour condensing through cooling.

Weathering and erosion in deserts

The forms of weathering and erosion that we have seen up till now, always made use of water in one way or another: frost action, solution of rock, transport by means of rivers or rainwater. In deserts there is not much water available for weathering and erosion, yet weathering and erosion do occur although not as rapidly as in less arid regions.
The main type of weathering in deserts is physical weathering. As in other environments, weathering starts with small faults and splits in the rock as a result of pressure reduction. The real erosion of the rock into stones and grains occurs due to the growth of salt crystals. Salts brought in by wind, mist and rain, precipitate in the small faults and splits in the rock. The salts form crystals which, as they grow, widen these faults and splits until the rock breaks, as in the role of ice crystals in frost action. The high temperatures in desert regions cause water to evaporate quickly, enabling the salt crystals to grow. The weathered material that arises from this process is not transported away by rivers or rain water. If rivers do arise in desert regions, for example through sporadic cloudbursts, then they do not

The dominant form of weathering in deserts is physical weathering. Here too, weathering begins with faults and cracks as a result of pressure reduction. The true disintegration of the rock into larger and smaller fragments is through the growth of salt crystals. Salt crystals transported by the wind, or in mist and rain, can precipitate into the small faults and cracks in the rock. The salts form crystals which, when they grow, push the faults and cracks a little wider apart each time, until the rock disintegrates.

The name, desert, does not apply to a region with a specific appearance, but to a certain climatic type. Deserts are regions where the annual potential evaporation is higher than the total annual precipitation or where the total annual precipitation is less than 250 millimetres. A desert's appearance is thus principally determined by the original geological structure of the region.

tend to flow for a very great distance. In contrast to rivers in other regions, the volume of water carried by a river through a desert region tends to reduce rapidly the farther downstream it travels due to evaporation and infiltration. Wind is the chief transporter of weathered material in deserts. The high surface temperatures of the land mean that the air above the desert quickly heats up, causing it to rise quickly. Cooler air from surrounding areas must then be drawn in rapidly to replace this warm air, often causing a strong wind to blow. However, the capacity of the wind to move material is much less than that of water. At normal wind speeds (not hurricane force!) only smaller particles are able to be transported.

The lack of water available for weathering and erosion, means that these processes are much slower in desert regions than in less arid regions. Historically interesting utensils or remnants of ancient civilisations are often extremely well preserved as a result.

Even in desert regions, weathering is at its most prominent on the faults and cracks in the solid rock. The growth of salt crystals widens and deepens them, creating pillars. Often the pillars are created under a layer of weathered material, which when it is removed by the wind, exposes the pillars to the sandblasting action of the wind.

Fraser Island, Australia, the world's largest sand island. Apart from a few sandstone rocks like 'The Pinnacles', the island is composed entirely of loose sand. Following a shipwreck here in 1836, the island was named after one of the survivors, Eliza Fraser, the wife of the captain.

Sand deserts

The precise appearance of a desert depends, to a great extent, on the original structure of the region. Only a few of the deserts in the world are actually sand deserts, vast endless plains with nothing more than dunes and fata morganas. About one third of the Arabian peninsula, the most sandy of the desert regions and one fifth of the Sahara, the largest desert in the world, are actually sand deserts. Sand plains are generally found in the lower parts of deserts, to where the weathered material, if it is fine enough, is carried and then deposited by the wind. This material usually originates from mountains and plateaus at the edge of the desert.

The lack of vegetation allows a continuous stream of fine particles to be moved by the wind. The smallest particles remain in suspension in the air a few centimetres above the ground. The slightly larger particles are transported a little further each time and collide with other particles which are then forced into the air.

In regions with large quantities of fine-grained sand, dunes are formed. If the wind flow is disturbed by an obstacle such as a large rock, then in the lee (shelter) of the rock, the wind speed reduces and the wind will lose some of its sand particle load. Hence this obstacle grows in size, disturbing the wind's flow more and more, and causing the wind to drop even more material. In this way sand dunes can sometimes reach heights of 500 metres. Dune heights of between 30 and 100 metres are quite normal.

Different sorts of sand dunes result from variations in the amount and size of the sand grains, the wind speed and the prevailing wind direction. The most common form of sand dune is the barchan, or crescent dune, which is crescent-shaped with the 'horns' pointing downwind. The manner in which dunes are formed means that not only do they migrate but they also increase

in size. A typical barchan dune will annually move downwind by some 10 to 30 metres.
Transverse dunes are dunes with long, straight ridges that lie transverse to the wind. These dunes are often the result of wave-like air currents over a flat surface, but they can also occur if several barchan dunes merge into one another. These dunes also migrate in the direction of the wind.
Longitudinal dunes have a very similar appearance to transverse dunes, but they lie parallel to the wind direction and are therefore formed in a totally different way. These dunes can be many tens of kilometres long, even extending to more than 100 kilometres.
All these various dune types have much the same internal construction. As can be seen in the photographs, a dune usually has a gently sloping side and a steeply sloping side. The gently sloping side lies to windward, where sand grains are continuously being deposited on it by the wind. Larger grains remain at the foot of the slope, with smaller grains migrating up the slope

If there is sufficient sand then the dunes steadily grow in height and come to lie on top of each other. The form of a buried dune is preserved under the new dunes. Dunes at the bottom of the pile will slowly become more compact and more solid. This allows them to be preserved for a very long time. If later weathering and erosion expose a section, then the structures of the buried dunes can still be quite clearly seen.

towards the top and then either lying there or toppling over the edge. The gentle slope does not only grow in height but also in the direction of the wind. The grains that remain close to the top of the dune cause the upper edge of the dune to be unstable and time and again the leeward edge collapses along the steeper side. In this way, the dune migrates in the direction of the wind.
If there is a sufficient supply of sand then dunes will grow steadily in height, as well as growing over one another. The structure of a buried dune is

preserved under new dunes and if sufficient sand and new dunes are deposited on top, then the buried dunes will slowly become more compact and solid. In this way they can be preserved for a very long time. The internal structure of dunes, laid down millions of years ago and which have now

Here and there in extensive deserts, isolated mountains protrude out of the desert surface. These are so-called inselbergs, meaning 'island mountains'. They predominantly occur in shield deserts, ancient flat regions which are characterised by plateaus and gravel plains. Ayers Rock (3.6 kilometres long and 348 metres high), Australia.

been re-exposed through weathering and erosion, demonstrates once again that the current processes involved in rock building, also occurred in exactly the same way aeons ago. Many of the sand deserts present nowadays were formed during the last ice ages. At that time, the continents then lay in the same positions as they now occupy. Large sections of the world were covered with ice as a result of the dramatic fall in temperature. In most desert regions ice never formed, but the climate was more arid and the wind stronger than at present, thus allowing large quantities of sand to be transported at that time.

Shield deserts

Large sections of the earth's deserts consist of plateaus and gravel plains, characteristic of shield deserts. The name shield reflects the way in which that part of the earth's crust arose geologically and is not a reflection of the type of desert that lies on it at present. Shields are the oldest regions of continents, to which the remainder of the continent has become attached over time. Thus shields are stable regions far from tectonic plate boundaries

Ayers Rock, Australia.
Whilst 'normal' mountains are generally the result of elevation with respect to their surroundings, inselbergs are huge blocks of stone that have been left over after weathering and erosion of the surrounding rocks.
They consist of a large block of very resistant rock that weathering has not been able to attack. Water from incidental rain showers runs off rapidly and concentrates weathering even more on the surrounding rocks.

Detail of the inselberg of Ayers Rock in Australia. This massive block of stone has almost no faults or even cracks; weathering therefore only takes place on the outermost layer. The rock is disintegrating very slowly, tiny flake by tiny flake. This process takes much longer than the disintegration of the surrounding, less resistant rock so that the mountain sticks out above the flat countryside.

Devils Marbles in central Australia. These large granite balls were originally part of a massive granite intrusion. Their present rounded shape is entirely the result of the very slow weathering of the hard rock.

Just like Ayers Rock, these granite balls are solid rock and free of any faults. Weathering here is a process of transforming minerals. The new minerals generally have a larger volume than the minerals that they replace. The result is that the rock is being peeled like an onion. Weathering acts on the entire surface area and larger and smaller fragments peel off. This process is called exfoliation.

and consist of very ancient rocks. Shield deserts are, on the whole, fairly flat with the ancient rocks lying beneath a layer of weathered material which, in addition to sand, also contains rock fragments.
The important role played by the wind in desert formation can be clearly seen in this type of desert. Stones that protrude from the surface or are partly buried are smoothed by the sandblasting action of the wind. If the wind blows from a particular direction most of the time, these stones acquire, over the course of time, a wind-cut facet on the windward side. These are called ventifacts.
Another typical result of wind action is the desert pavement. Stones and rock fragments that were originally uniformly distributed throughout the thick layer of weathered material, become concentrated on the land surface as the finer particles and granules are removed by the wind. The desert surface becomes 'paved' with smaller and larger stones.

Kings Canyon, a 200 metres deep river valley in central Australia. This warm, arid mountain region is completely surrounded by deserts.

In places in the vast shield deserts, isolated mountains rise high above the flat landscape. These are called inselbergs, which means 'island mountains' in German. If normal mountains are the result of elevation in respect of their surroundings, inselbergs are large rocks left over after weathering and erosion have taken their toll of the surrounding rock. They consist of a large block of very resistant rock that proves very difficult to weather. Water from

the sporadic rain showers rapidly runs off this resistant block and further concentrates weathering on the surrounding rocks.

Large sections of the deserts on the earth consist of plateaus and gravel plains, typical of shield deserts. The name, shield, alludes to the geological origin of a piece of the earth's crust and not to the desert that is present on it. Shields are the most ancient parts of the continents, to which the rest of the continent has become attached. They are now very stable regions, far from the tectonic plate boundaries and consist of very ancient rocks. Shield deserts are generally very flat with inselbergs occasionally rising above the flat landscape. The Olgas, Australia.

Some 90% of the earth's crust consists of igneous rocks. After a few billion years of weathering and erosion of igneous rocks, the earth's surface (thus not the entire crust) now consists of 75% sediments and sedimentary rocks.

The sediments on the earth's surface pass through the geological cycle and in old rocks which have been attacked by weathering and erosion, structures can often be recognised which are, at present, being formed in other parts of the earth.

6 Caves

This last chapter looks at products of weathering and erosion that, apart from being beautiful and impressive, have, throughout the centuries, been used functionally as shelter by both animals and man. In many places in the world, prehistoric paintings bear witness to the earlier presence of man. In later times, man has frequently made use of these naturally constructed shelters or hiding places, just as many animals still do today. These shelters are of course, small caves on the surface of the earth, recesses in mountains and hills, for instance either in granite or sandstone. The action of frost and the growth of salt crystals in weak zones in the rock allows weathering to penetrate deep into the rock.
Even more beautiful and impressive, but less functional, are caves, containing stalagmites and stalactites; these are hollowed out in limestone regions by chemical weathering. These caves lie below ground level and often consist of a complex system of chambers and passages that may be many kilometres long. Limestone caves can occur in regions which have a thick layer of limestone as the bedrock, where chemical weathering is highly advantaged compared to other sorts of weathering. High temperatures, plentiful precipitation and vegetation, and a high ground water level are all important criteria in cave formation. The carbon dioxide absorbed from the atmosphere and rotting vegetation provides a weakly acidic ground water, which can easily dissolve calcium carbonate, the main component of limestone.
Slowly moving ground water widens faults, splits and pores. The wider any opening, the more water can flow through this and the greater the quantity of limestone that can be dissolved. Any such opening rapidly grows wide enough to become a cave. The fascinating stalagmites and stalactites in the cave only

Throughout the ages, both men and animals have made frequent use of naturally formed shelters and hiding places. These are mainly small caves on the surface of the earth in, for example, granite or sandstone, which are the result of weathering through frost action or the growth of salt crystals in weak zones deep in the rock. Crete.

In this cave in Slovenia, the stalactites have formed a stone curtain.

The many colours come from impurities in the calcium carbonate. Pure calcium carbonate is white, but if, for instance, it contains iron particles then the limestone will be coloured yellow and red by iron oxide.

▷Stalactites, stalagmites, curtains and columns in the Cango Caves in South Africa, a kilometre-long complex of caves. In addition to stalactites and stalagmites, other structures can also be found in limestone caves, such as, for example, calcite bubbles, thin films of limestone that form over puddles of water and crystal needles of gypsum.

start to form once the water table has dropped and the cave is filled with air. Water, permeating through the limestone from above forms droplets on the ceiling of the cave. The flow of air through the cave causes some of the carbon dioxide gas in these water droplets to evaporate. In a chemical reaction this is the opposite to the solution of limestone, a small amount of the calcium carbonate dissolved in the water then precipitates out on the ceiling. Many, many, drops of water later a stalactite begins to form on the ceiling. Drops that fall from the ceiling or from a stalactite also leave a deposit of calcium carbonate on the floor of the cave. This is how stalagmites are formed, and these may sometimes stand directly under a stalactite. If both stalactite and stalagmite continue to grow they may join up to form a single column.